KONG FLU PANDA 11: CHAOS AND PANDAEMONIUM

BY TYLER LAZARUS STUMP

AKA MISTER.E

PREFACE:

So, here is the deal.

I wrote an entire series for these books. . . called "KONG FLU PANDA."

They sold around the world.
They elevated life science, and virology up to the top of the charts.
They proved the genre can sell.
They did really well.

I swore I was done with them··· after book ten.

I still am.
···"but why are you writing an 11th, then?????"

Because I had no choice.

NEW and exciting things (bullshit things) happened that were all connected to the events of the first ten, regarding a pandemic, Trump, Biden, and the downfall of America.

Because of these events, I felt I had a duty (no choice) to record the latest chapter.

It also would have bothered me for the rest of my gay life to not capture these historical moments··· and leave the ending incomplete.

So⋯ I went back and re-opened "the case," for the sake of literary accuracy and posterity.

Contained below are more shocking, insane, off-the-wall and nutty parcels of history that really happened.

Whoever is reading this, far in the future, if there is any world and future LEFT to even inherit⋯ this is what happened.

I recorded it all.

Much love,

Tyler Lazarus Stump

MISTER. E

Donald Trump had been booked into jail,
fingerprinted, mugshotted,
set on $$$bail,
and released.

A president.
had been arrested.
and let out on 200k bond.

America was never going to be the same.

For many reasons.

1. Trump had been indicted on conspiracy to overturn an election,
submit fraudulent documents,
signing false forms,
and other counts of impersonating officials
to secure those forms.
PLUS...
A whole list of charges,
and multiple attorneys and OTHER people
were also booked and bonded
for helping commit the crimes.

2. While these crimes were being punished,
The Biden Family and Hunter Co. Crimes
weren't even being pursued.
..... $$$$$$$$$$$$$$$
While multi-MILLION dollar deals to China
had been secured through Hunter and Co,
Joe taking a percentage of those deals,
and Joe and Bro Weiner Bator
Making those deals with Chinese nationals...
WHILE JOE
was vice-president.

Leading up to present moment.
Of Donald Glaring in his mugshot.

Was Donald Guilty? Absolutely.
The law will also decide.
(Even though he was caught on audio recordings doing all of it)

But the problem is,

This wasn't about Trump.

Not Really.

It was about a dying America, A Banana Republic,
Democratic Hypocrisy,
A Nation imbalanced in its pursuit of "law and order,"
Ignoring the crimes of the other (Joe Joe and Hunter)
And allowing political games,
which had been being played for a decade to
TRUMP actual executive legitimacy.

It was about the long-standing decay, deterioration,
communist onboarding,
constitutional republic enamel eroding.

It was about a gnashing of teeth,
liberals and republicans foaming at the mouth,
and red and blue parties running their mouth,
as Biden and previous administrations
had ran the country
INTO THE GROUND.

It was about Afghanistan, Ukraine, Iraq and Iran,
It was about Bush starting false Wars,
It was about Obama being a wolf in sheep's clothing,
It was about Reagan lying,
It was about Bush Sr. being friends with the Osama Bin Laden
Family,
It was about Oil,
Endless (Forever) Wars,
And Biden and Trump
just being the farewell tour
of a long-forgotten legacy act.

It was about the end of America.

The Decline.

The ICU-patient coding and seizing

on the operating table.

The Stage 4 cancer patient gasping out their last breaths, before fading away.

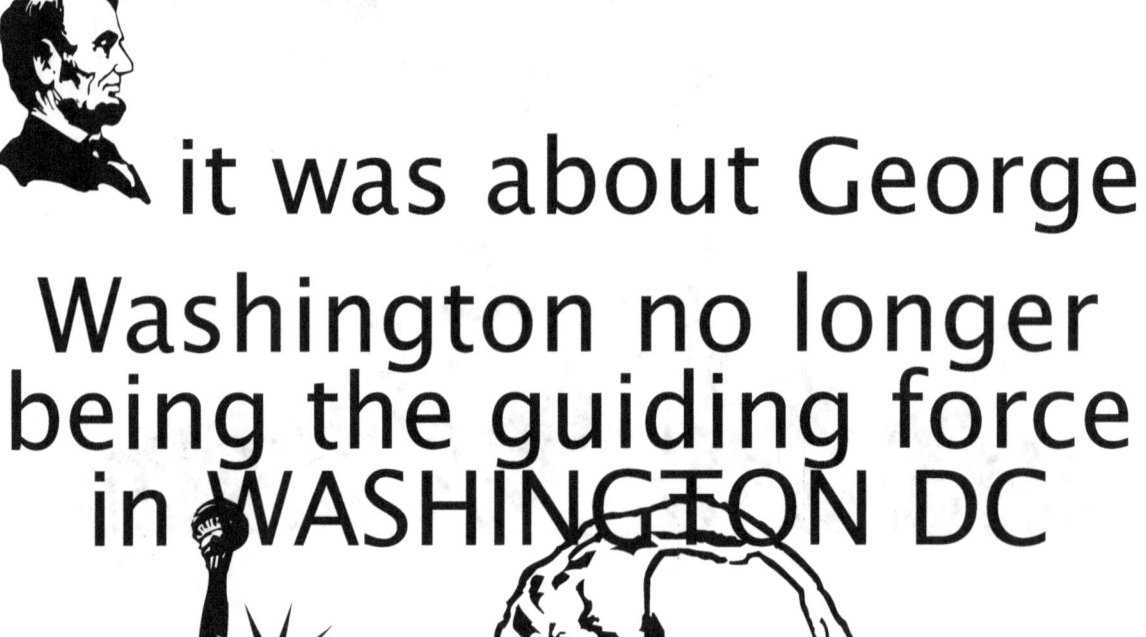it was about George Washington no longer being the guiding force in WASHINGTON DC

It's about a fractured country.
A broken country.

A country past the point of saving.

Trump being trash,
was just a symptom of a country that had been turned into a
trashcan.

See Cocaine Being Found in the Whitehouse,
During Biden's admin,
on the week Hunter visited.

It was about white lines of coke,
and black lines of history being made.

The country had entered into a death spiral.

And a spiral of lies,
reached all the way to Trump, Biden,
Wuhan,
Russia,
UKRAINE,
China,
And To the World Economic Forum
and back again.

Lies which had gotten people killed.

A virus which murdered 25 million.
A vaccine which didn't vaccinate and has killed likely more.

Trump going to jail was the least of your problems.
You have Daddy Pfizer,
Moderna,

and Johnson and Johnson

Slapping you with their Pharma Johnsons,
and giving you blood clots
when you finish.

But you also Have a China INTENT on securing the #1 spot,
reforging alliances with local nations on their economic pact,
rewriting history, and holding hands with Russia,
AS russia was threatening nuclear war.

Don't worry about Trump Stealing elections,
you won't have a country worth a damn to even keep going,
at this rate.

As Trump puts his finger prints into Fulton County Jail,
Biden praises the rise of China, after making deals with Chinese
Communist officials, through his son,

as China surges to the top, backs Russia,

and America collapses and continues funding/funneling billions of dollars into a corrupt Ukraine,
who has some kind of blackmail material on American Officials.

And communist/soviet forces advance,
as American forces begin to self-destruct.

All of this is happening after and during an engineered virus + bioweapon killing, a vaccine murdering, ukraine mooching, China, Xi, Putin and Co advancing on an America caught arguing about Trump and Biden,
as both Trump and Biden commit clear treason against their own countries.

So much is going on, ALL AT ONCE, I may have to do some Fung-Shi, and consult the I-CHING.

pushes my eggroll around and eats my fried rice

MANY FIRES, HAPPENING IN MANY DIFFERENT POLITICAL PLACES.

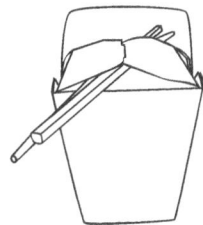

But there really is no reason stressing out about it,
because no empire lasts forever.

The American Arrogance to think it could go on forever,
as it fell apart internally is JUST another
CLASSIC
case scenario of what happens when governments become
corrupt,
leaders become evil,
a populace becomes fragmented, polarized, apathetic and
indifferent,
and what happens when a stupid populace
elects stupid people
to lead them.

**It leads only to trouble, tears,
chaos and pandaemonium.**

HAD A STUPID AMERICAN PUBLIC NOT VOTED FOR STU*PID AMERICAN TRAITORS...

They wouldn't have had one former president taking a mugshot photo,
They wouldn't have a current one struggling with dementia and selling out our country to China,

They wouldn't have had the cost of living raised so exponentially that everyone was struggling and BECOMING POOR,
they wouldn't have praised "The rise of China,"
and had PRESIDENTS celebrating communistic expansion,

They also wouldn't have praised Russian dictators,

And they wouldn't be getting taxed to the DICKENS.

But that's not what happened,

Because Stupid Americans, argued about STUPID SHIT (trannys, guns and abortion)

and elected stupid fucking men that never should have been president.

And now,

<u>America had been turned into the land of the stupid.</u>

TERRIBLE POLICY CHOICES

HAVE TERRIBLE END CONSEQUENCES.

■

So No,
I did not care or give a sh*it about Trump,
Trump going to Jail,
Because I couldn't stand TRUMP OR BIDEN,
BECAUSE NEITHER WERE GOOD
HAD BEEN GOOD FOR
OR WERE EVER GOING TO BE GOOD
FOR MY COUNTRY.

**TRASH MEN
HAD TRASHED THE PLACE.**

MEAN EFFING WHILE,

CHINA AND RUSSIA GET EVERYTHING
THEY WANT,
AMERICA FALLS TO THE DOGS,
MODERNA AND PFIZER GET AWAY
WITH ABSOLUTE INSANITY,

WUHAN GETS OFF GUILT FREE,
THE NIH FADES TO THE SHADOWS
(AFTER FUNDING THE LAB IN
WUHAN)

AND ALL I HEAR ABOUT ARE
STUPID< SELF ABSORBED USELESS
CELEBRITIES
PROMOTING THEIR NEXT STUPID
SELF ABSORBED PRODUCT AND
ALBUM.

Oh, and Trump got caught stealing classified documents. In his bathroom. Biden was hiding classified documents IN HIS GARAGE.

I don't believe in God, But GOD BLESS.... America.

WHAT A MESSY ASS COUNTRY.

You all stole this place from Indians, and Chief BigThickPole7inches

Cursed YOU ALL.

THE END.

To the background of this report,
beyond the continued and what feels like, eternal nonsense of Biden and Trump... a Chinese Lab conducting illegal experiments on Covid, HIV, and other viral specimens was found in California; operating totally unchecked.

A biolab, or more promptly, a bioweapons lab.

A call came in an old warehouse that was illegally using someone else's water supply hose, code enforcers got there, and found a ventilation fan blowing foul air from the back, and then checked to see the permit of the building. There is none.

They get inside, and find dozens of refrigerators, with blood, chemicals, lab mice (ACE-2 transgenic mice, I'll explain why these lab rats are so important further down) + a plethora of illegal lab supplies and virus samples.

Beyond COVID19, they were also housing Hepatitis, Malaria, Chlamydia, Herpes and other viral strains.
Lovely.
Owned by a Chinese company, under Prestige Biotech INC.

Along with the over 1,000 lab mice, they also had jars of Urine and animal specimens being kept in crowded, soil containers.

Major News Outlets Pick up the story,
USA TODAY: "Fairly Shocking: Secret Medical Lab in California stored Bioengineered Mice laden with Covid."

NBC NEWS: "CDC detects Coronavirus, HIV, Hepatitis, and herpes at Unlicensed California Lab."

NEWSNATION: "Biowarfare concerns surface with discovery of illegal Calif. Lab."

Washington Times: "House panel Launches investigation into illegal chinese lab."

Here's the deal.

People (THE CDC) tried to deflect saying it "was no big deal,"
— it's a huge deal — and downplay it as saying it's just "anti-asian sentiment."

First, this has nothing to do with Chinese people.
But it does have to do with illegal chinese labs, backed by

the communist party.

Remove race from this,
and place the human Race back in,
and realize you have nefarious entities engaging in illegal
activity (highly illegal) inside your own backyard.

It's now a race to the end,
as Communism has rooted itself inside your universities,
and is hiding in laboratories.
With samples, specimens, and blood.

And denying that is tantamount to having blood on your
hands.

Two, operating biolabs without any kind of registration or
biosecurity clearance PLUS operating as a medical lab, is
not just totally wrong, it's 100% illegal. So, shove that
nonsense up your ass.

Wrong IS wrong.
ON A FEDERAL AND STATE LEVEL.

We wouldn't tolerate that for our own citizens,
but yet we should be making unrestricted accommodations
for COMMUNIST nations????????

EXCUSE THE F.UCK. out of ME.

HELL NO.

Those unrestricted "accommodations"
are going to lead to unrestricted Bioweapons.

Which leads me back to the importance of them noting how
many ace-2 transgenic mice the facility was using.

Experimenting on.

Those mice aren't ordinary house/field mice.
They are bred and created with human cells/tissue lines to
experiment with medical products to see how it will
potentially fare in ACTUAL human subjects.

But there's more, much more to the relation to COVID19,
COVID variants, and the creation of COVID19.

This is straight from PNAS.org
(The Proceedings of the National Acad. of the Sciences of
the USA)

It's the most prestigious and reputable, and cited scientific journal out there.

With federal and military applications.

It's THE BIG one.

I'm taking from an article, published in 2008.

"Synthetic recombinant bat SARS-like Coronavirus is infectious in cultured cells and in mice."

Approved October 14, 2008
https://DOI.org/10.1073/pnas.0808116105

SARS-COV Mouse-Adapted Spike Mutation Enhances Bat-SRBD Replication in Mice.

The study goes onto show the importance of using the ACE-2 mice to infect the murine (mouse) tissues and cultures, and what ACE-2 mice used, and how they used them to foster the infection.

SO, fast forwarding to present day - the hell that is 2-0-2-3 - Finding a massive supply of live ACE-2 mice with covid-19 samples and coronavirus specimens shows (conclusively, incontrovertibly) that what they were doing was malicious, and as malignant as it gets.

Those mice only have one purpose in this context, to act as

a medium to creating and experimenting with past and future viral strains of this specific virus.

They aren't there just as "mice,"
they are very specific lab rats.

From this specific, illegal, highly shocking chinese communist lab.

Operating ILLEGALLY, secretly inside THE USA.

And even worse, the authors of that study ended up later working as direct collaborators with the Wuhan Institute of Virology.

I'm not kidding.

But it gets worse, kids.

When Omicron first released and appeared, scientists couldn't figure out where it came from. For a host of reasons.
Including the changes of how it attacked its hosts.
1. It had so many unique nucleotide changes to the viral architecture that wouldn't be usual or normal for typical viral mutation cycles.
2. It was so infectious it covered the globe, beyond Alpha, Delta and ORIGINAL WUHAN STRAINS, within 3 months.

Basically, anyone who caught covid between 2021-2022, had been hit with OMICRON VARIANT.

And then, upon genomic sequencing, they concluded:

"Evidence for a Mouse Origin of the SARS-COV-2 Omicron Variant." - J. Genet Genomics, 2021 December

"Did Omicron come from mice??" - The New Yorker

"COVID19: Did Omicron evolve in Mice?" - Medical News Today

"Evidence for a Murine Origin of Omicron." - ScienceDirect

"If Omicron does have a mouse origin, then what could be next?" - Infectious Disease Society of America.

It shows 100% proof of being passed and evolved through a murine system, but with directed and not natural evolution.

AKA LAB MADE.

So, now you see why finding random ACE-2 mice with COVID samples in an illegal lab inside America is beyond a code violation or medical biohazard.

It's 1000000% the prelude of a bioweapon attack against

human biological systems.

Those mutations found inside this variant couldn't be explained and the changes to the spike protein were profound.

In such a way that natural mutations couldn't make this happen.

And I'm not making this up.

You're under attack.

You were attacked.

And you still are.

Cheerio.

www.ingramcontent.com/pod-product-compliance
Lightning Source LLC
Chambersburg PA
CBHW080925290526
45795CB00007BA/2654